Yaya Sagnon
Fidèle Zongo

Phytoremediation of Ouagadougou landfill soils

Yaya Sagnon
Fidèle Zongo

Phytoremediation of Ouagadougou landfill soils

Agronomic and environmental valorization of anarchic landfill soils in the city of Ouagadougou

ScienciaScripts

Cover image: www.ingimage.com

This book is a translation from the original published under ISBN 978-620-2-54079-7.

Publisher:
Sciencia Scripts
is a trademark of
Dodo Books Indian Ocean Ltd. and OmniScriptum S.R.L publishing group

120 High Road, East Finchley, London, N2 9ED, United Kingdom
Str. Armeneasca 28/1, office 1, Chisinau MD-2012, Republic of Moldova, Europe
Managing Directors: Ieva Konstantinova, Victoria Ursu
info@omniscriptum.com

Printed at: see last page
ISBN: 978-620-8-59597-5

Table of contents

DEDICATION

I dedicate this work to my mother and father for the education they gave me, for the sacrifices they made, for their unconditional support,

To my grandmother for her blessings,

To the SAGNON, BELE, BIHOUN, TOU families and dearest friends for their immeasurable support.

ACKNOWLEDGEMENTS

Above all, we thank Almighty God for giving us the strength, will, courage and patience to complete this work. We would also like to express our sincere thanks to :

- To **Prof. Bila Gérard SEGDA,** Director of the Institut de Génie de l'Environnement et du Développement Durable (IGEDD):
- To **Prof. Salifou K. OUIMINGA,** Deputy Director of IGEDD :
- To **Dr S. Zacharie KAM,** Head of the IGEDD Department
- To the **IRD Representation in Burkina Faso**, for access to the infrastructure that enabled us to carry out our work;
- To **Prof. Edmond HIEN**, Professor of Agro-Pedology, Joseph KI-ZERBO University, Director of the Soils, Materials and Environment Laboratory, our thesis supervisor, who, despite his multiple occupations, accepted the responsibility of supervising this work;
- To **Dr Koulibi Fidèle ZONGO** , Enseignant-Chercheur au Centre Universitaire de Tenkodogo, Agro-pédologue, our dissertation Co-Director, who spared no effort in terms of availability, guidance and financial support. Thank you for your unfailing advice and support. Your rigor is undoubtedly an example to us all;
- To **Dr Daouda GUEBRE** for his support in the smooth running of our work;
- To **Mr. Prosper SAWADOGO,** laboratory technician at IRD's UMR Eco et sols, who greatly contributed to the success of this work;
- To **Dr Telado Luc BAMBARA**, Enseignant-Chercheur at the Ecole Normale Supérieure de Koudougou, for his scientific contributions to improving the quality of this document;
- To my cousin, **félicité BIHOUN**, certified high school and college teacher, for agreeing to proofread the document and correct grammatical errors;
- To all the IRD staff for their warm welcome and support during our stay;

My thanks also go to the students of the Master's program, class of 2021-2022, particularly those in the Territory-Environment-Health option, with whom we have been able to form a family, and to all the teaching staff at IGEDD.

Last but not least, we would like to thank all the people who contributed to this work.

SUMMARY

The aim of the present study was to assess the effect of soil from various landfill sites on agronomic parameters, and the ability of lettuce and amaranth plants to accumulate or eliminate Cu, Pb and Zn from contaminated soil. The methodology consisted in testing six (06) treatments with five (5) replicates in a greenhouse for 75 days:

Toudouwéogo landfill soil (under amaranth cultivation); Saaba landfill soil (under amaranth cultivation); absolute control (under amaranth cultivation); Toudouwéogo landfill soil (under lettuce cultivation); Saaba landfill soil (under lettuce cultivation); absolute control (under lettuce cultivation).

The physical and chemical compositions and trace metal (TME) contents of the 2 landfills were studied before sowing. TMEs in landfill soils and plants at harvest were also determined. Morphological parameters and plant biomass were measured. Atomic Absorption Spectrometry (AAS) used to determine MTE content in soil samples and biomass. Analyses of variance (ANOVA) and TukeyHSD multiple comparisons of means were performed on the agronomic data at the 5% probability threshold ($p < 0.05$). The results showed that the soils of the Saaba and Toudouwéogo landfills exhibit variability in physical, chemical and TME content parameters. Agronomically, the landfill soils significantly improved the morphological performance and biomass production of lettuce and amaranth, with the Toudouwéogo landfill soils performing better agronomically than the control soil. In terms of phytoremediation, results showed that TME concentrations (Cu, Pb, Zn) on all landfill soils were well above limit values. The pollution index for each soil was greater than 1, i.e. 8.15 for the Saaba landfill soil, and 7.50 for the Toudouwéogo landfill soil. Plant TME concentrations would appear to be influenced by soil levels. Zinc was the element most accumulated in the plant biomass, having a very high content in the soil compared with the other TMEs studied. The two plant species contributed to a reduction in the soil pollution index at harvest of 1.6 for the Saaba landfill and 6 for the Toudouwéogo landfill, but the values still exceeded the standards for agricultural use.

Keywords: landfill, soil, amaranth, lettuce, agronomic performance, TME, pollution, phytoremediation.

ABSTRACT

The present study aimed to evaluate on the one hand the effect of soils from different public landfills on agronomic parameters and on the other hand, the sanitation capacity of lettuce and amaranth plants to accumulate or eliminate Cu, Pb and Zn from contaminated soil. The methodology consisted of testing in a greenhouse for 75 days, six (06) treatments with five (5) repetitions : Soil from Toudouwéogo landfills (under amaranth cultivation) ; soil from Saaba landfills (under amaranth cultivation) ; absolute control (under amaranth culture) ; soil from the Toudouwéogo landfills (under lettuce cultivation) ; soil from Saaba landfills (under lettuce cultivation) ; absolute control (under lettuce culture). The physical and chemical compositions and trace metal element (TME) contents of the two landfills were studied before sowing. TMEs in soils and plants from landfills to harvest were also determined. The morphological parameters and the plant biomasses produced were measured. The determination of ETM contents of soil samples and biomass was carried out by Atomic Absorption Spectrometry (AAS). Analyzes of variance (ANOVA) and multiple comparisons of TukeyHSD means were carried out on the agronomic data at the 5% probability threshold ($p < 0.05$). The results showed that the soils of the Saaba and Toudouwéogo landfills present variability in physical, chemical and ETM content parameters. On the agronomic level, the soils of the public landfills significantly improved the morphological performance, the production of biomass of lettuce and amaranth with a better agronomic performance of the soils of the Toudouwéogo landfills compared to the control soil. In terms of phytoremediation, the results showed that the concentrations of ETM (Cu, Pb, Zn) on all landfill soils were well above the limit values. The pollution index of each soil was greater than 1, i.e. 8.15 for the Saaba landfill soil, and 7.50 for the Toudouwéogo landfill soil. ETM concentrations in plants would be influenced by their levels in the soil. Zinc, which had a very high content in the soil compared to the other ETMs studied, was the element most accumulated in the plant biomass. The two plant species contributed to the reduction in the soil pollution index at harvest of 1.6 for the Saaba landfill and 6 for Toudouwéogo soil but the values still remain above the standards for use in agriculture.

Keywords : public landfills, soils, amaranth, lettuce, agronomic performance, ETM, pollution, phytoremediation

INTRODUCTION

Urban and peri-urban market gardening is on the increase in large African cities in general, and Ouagadougou, the capital of Burkina Faso, and its suburbs are no exception. Urban market gardening in Africa is seen as a solution to the problem of supplying vegetables to increasingly populous cities [1]. Population growth and urban industrialization generate waste that can contaminate the environment [2], disrupting ecosystems and affecting living beings. This waste, whether of agricultural, municipal or domestic origin, can constitute a quality organic amendment for soils, rather than being stockpiled in public landfills where its decomposition emits greenhouse gases [3]. To satisfy ever-increasing demand, market gardeners adopt a strategy of improving their productivity, based on the use of chemical fertilizers and pesticides [4]. Indeed, some farmers, lacking the means to buy chemical fertilizers (too costly and not readily available), go to urban landfill sites close to their fields, where they may or may not burn household waste and recover the debris to amend the soil before growing vegetables. While this practice contributes to increased productivity through the addition of organic matter, it also results in metal contamination of market garden soils and crops. One of the main ways in which humans become impregnated with these heavy metals is through the consumption of plants grown on these contaminated soils. At certain concentrations, these heavy metals are known to have harmful effects on consumer health [5].

One solution to this problem is to treat landfill soils to remove metallic elements, or at least reduce their concentration to acceptable levels. One such treatment method, which emerged in the 1990s, involves the use of plants capable of growing on soils with a high metal content and likely to mobilize, or absorb, significant quantities of these metal pollutants [6]. This technique, known as phytoremediation, is proving promising because it is both inexpensive and more respectful of the environment [7]. And these are two of the reasons why it is so important to take an interest in a study of the reclamation of landfill soils that can be used to amend agricultural soils.

The overall aim of this study is to assess the effects of anarchic landfill soils on amaranth and lettuce production, and the phytoremediation of these two plants in heavy metals.

Specifically, it involves

- Characterizing anarchic landfills in the city of Ouagadougou;
- Evaluating the effect of landfill soils on the agronomic performance of amaranth and lettuce;
- Evaluate the efficacy of phytoremediation of lettuce and amaranth.

Working hypotheses were formulated:

- Ouagadougou's uncontrolled landfills contain organic substrates and mineral compounds, as well as pollutants, particularly trace metals (TMEs)
- Urban waste affects the growth and plant biomass of amaranth and lettuce;
- Lettuce and amaranth can help clean up landfill soil.

I. BIBLIOGRAPHIC SUMMARY

1.1.A few definitions

Waste: Any residue from a production, transformation or use process, any substance, material, product or, more generally, any movable asset abandoned or intended for abandonment by its holder [8].

Uncontrolled (Anarchic) **landfill**: This is an uncontrolled open-air dump, poorly exploited, which receives mixed waste of different kinds: agricultural, commercial, industrial, domestic and hospital, and without prior treatment [9].

Controlled landfill: This is a landfill that has been approved and authorized to operate on an appropriate site according to a development plan, where the spreading of waste is controlled. These are transfer centers equipped with refuse bins for collecting and transporting waste to treatment and recovery centers.

Heavy metals: Heavy metals are defined as metallic elements with a density greater than 5 g/cm^3 and are most often present in the environment in trace form at concentrations that vary according to the environment and organism [10]. Heavy metals include major elements and trace elements, i.e. elements whose concentration in the earth's crust is less than 1% for each: mercury, lead, cadmium, copper, arsenic, nickel, zinc, cobalt, manganese and others. However, some toxic trace elements are not metals (arsenic, selenium). That's why it's better to use the general term trace metals (when they are indeed metals) or TMEs [11].

1.2 General information on heavy metals

1.2.1. Heavy metal sources

Heavy metals are naturally present in very small quantities in all environmental compartments. Heavy metals are said to be present "in trace amounts". In fact, human activity has brought no change in the quantities of metals in the environment, as they are non-biodegradable. Rather, it has changed the way metals are dispersed, their chemical forms (speciation) and concentrations, through the introduction of new modes of dispersion (smoke, sewage, various types of waste, cars, industry, etc.).

1.2.2. Heavy metals in municipal waste

Heavy metals are found in all waste categories: paper, plastics, batteries (mercury, lead and cadmium), overcaps and accumulators (lead). Batteries are sources of several metals, including Cd, Ni, Pb and Zn [12]. Work carried out on urban waste used as compost feedstock has shown that zinc is mainly found in the fraction bound to iron and manganese oxides, lead in the residual fraction and copper in the fraction bound to organic matter [13].

1.2.3. Heavy metals in soil

In soils, TMEs may have a natural origin, stemming from the initial chemical composition inherited from the geological material from which the soil is derived. The natural pedogeochemical concentration results from this condition [14]. In addition to the initial pedogeological inheritance, other types of MTE input can also occur. These include atmospheric deposition of anthropogenic origin (industrial dust), agricultural inputs (inorganic fertilizers, manures, pesticides), pesticides (Cu, Zinc, Fe, Mn) and waste applications (agricultural waste, composts, sewage) [15].

1.3 Impacts of heavy metals

1.3.1. On health

As heavy metals are non-biodegradable, their presence in the environment causes them to bioaccumulate and end up in the food chain through biomagnification, thus creating public health problems. Indeed, the accumulation of heavy metals in the body causes enormous adverse effects on human health, affecting all biological functions, even though there are no specific, detectable symptoms of metal poisoning. Among the most serious illnesses caused by metal intoxication, lead poisoning corresponds to Pb intoxication involving blood lead levels in excess of 100 ug L^{-1} of Pb in blood, leading to serious and irreversible effects on the body, notably mental retardation in children [16].

1.3.2. On plants

The toxic effect of TMEs on plant growth manifests itself in reduced growth of aerial and root parts, dramatically affecting biomass production [17]. However, it is important to note that not all plants are affected by the toxic effect of heavy metals with the same severity. Indeed, some so-called hyper-accumulative plants are able to grow, develop and reproduce in the presence of high concentrations of TMEs.

1.4. Treatment of polluted soil

1.4.1. Decontamination methods

The most common methods for decontaminating soil polluted by heavy metals involve excavating the soil for physical and thermal treatment. However, the main difficulties encountered are high costs, cumbersome implementation, alteration of the physical structure and disruption of the soil's biological activity.

1.4.2. Heat treatment

This method has been used for many years to clean up soil contaminated by organic pollutants and heavy metals. Contaminated soil is excavated and passed through the furnace of a flame reactor. The final product leaving the furnace is a non-leachable, glass-like slag. Volatile metals are captured in a dust collector, while non-volatile metals condense as a molten alloy.

1.4.3. Physical and chemical treatment

It consists in solidifying the polluted soil to improve its physical characteristics and reduce the pollutant's mobility in the soil. Certain mineral binders such as ash, lime and portland cement are required for solidification.

1.5. Phytoremediation

Phytoremediation is a method of soil decontamination using plants capable of fixing heavy metals, surviving and reproducing in conditions that are a priori hostile. It is an approach to environmental restoration using vascular plants that are hyper-accumulators or able to fix heavy metals contained in soils.

Phytoremediation falls into two main categories: phytostabilization and phytodecontamination.

1.5.1. Phytostabilization

It consists in immobilizing contaminants in soil and water in situ via plant roots by absorption, adsorption or precipitation in the rhizosphere [18]. This technique is applicable to sites where metal bioavailability (possibility of assimilation by plants) is reduced and pollutant movement is estimated. Unlike organic compounds, metals cannot be degraded and, consequently, effective clean-up requires their immobilization to reduce or eliminate their toxicity. [19]. This process will thus immobilize heavy metals present in the soil at the level of the plant root system, and limit the flow of pollutants to the depths of the soil by runoff, thus reaching groundwater.

1.5.2. Phytodecontamination

Phytoextration: This technique exploits the ability of certain accumulator or hyper-accumulator plants to transfer heavy metals from their roots to their leaf tissue via transporters, in order to extract pollutants from contaminated soil by storing them in the above-ground biomass. Heavy metals present in substrates are bioaccumulated via plant roots to their harvestable (usually aerial) part, which is then extracted from the environment by mowing. Harvesting is repeated until contaminant levels in the soil reach acceptable levels [18]. When it comes to processing the biomass produced, authors often refer to incineration and energy recovery. The ash, or bio-ore, will be confined to landfill sites as hazardous waste, or recycled in metallurgy.

Phytomining: Polluted biomass can be used to recover metals: this is known as phytomining. The metals obtained by phytomining are very pure and can be used directly as catalysts: this is known as eco-catalysis [20].

1.6. Effects of landfill soils on the agronomic performance of market garden crops

To improve soil fertility in market gardening areas, market gardeners generally use a variety of agricultural inputs, including animal waste, green waste, sludge and household refuse. The latter is usually taken from public rubbish dumps not far from the fields. Household waste, which contains the by-products of goat and sheep rearing, can have a positive impact on vegetable production, as it is a low-cost source of nutrients for market gardeners and can improve soil structure. It is therefore a source of fresh, transient organic matter that can contribute to soil fertilization. Animal matter is rich in fertilizing elements, while plant debris is a precursor to humus [21]. In Jos (Nigeria), soil amendment with ash from urban waste increased agricultural productivity [22]. The same observation was made in a study carried out in Nagoundéré (Cameroon), showing that the use of manure from municipal waste is beneficial for fertilizing market garden soils [23].

1.7 Effect of market garden crops on phytoremediation of landfill soils

Vegetable crops are an essential component of urban diets, rich in carbohydrates, proteins, minerals and vitamins. However, when vegetables are grown on polluted landfill soils, they can accumulate metals in high concentrations that can exceed standards. A study carried out on uncontrolled landfills in the city of Goma (Democratic Republic of Congo) showed that amaranth grown on uncontrolled landfill soil accumulated high levels of Pb and Zn compared with the maximum reference concentration [24]. Certain heavy metals, such as copper, zinc, manganese, iron and cobalt, accumulate in the soil and are absorbed by plants, which need them for growth and development. The amount accumulated by plants depends not only on vegetable-related factors, but also on soil-related factors. Similar observations were made by Veron on Cd accumulation by vegetables [25].

II. MATERIALS AND METHODS

1. Study environment and sampling site

1.1 Study environment

The trial was carried out in a greenhouse, in vegetation vases, at the Institut de recherche pour le développement (IRD) in the town of Ouagadougou. The greenhouse has the same climatic conditions as the city of Ouagadougou, which has a Sudano-Sahelian climate, with an average annual rainfall of 569 mm and an average temperature of 28.6°C.

1.2 Soil sampling sites for uncontrolled landfills

This study involved soil sampling at two anarchic landfill sites in the city of Ouagadougou. These were a landfill in Toudouwéogo and the landfill of a former brick quarry in the Saaba area. These two landfills and other anarchic public dumps were mapped using their geographic coordinates and ArcGIS version 10.8 software (Figure 1).

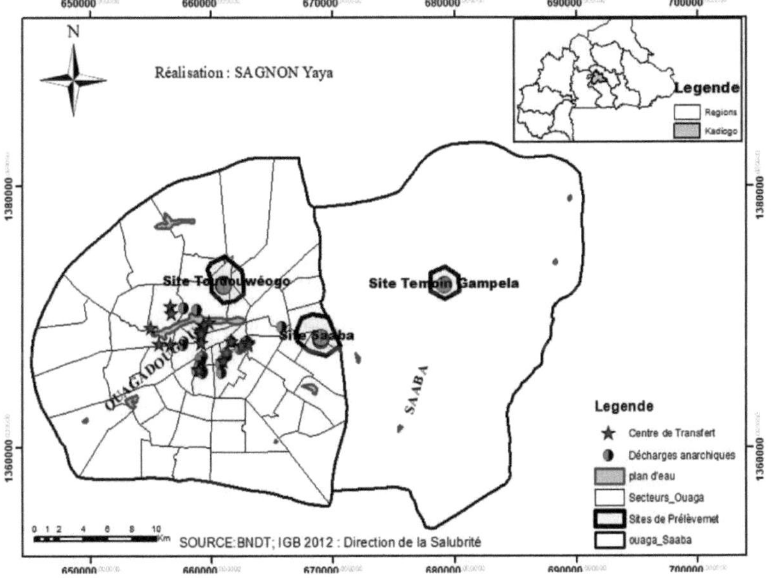

Figure1 : Location of soil sampling sites

2.plant material

Two plants were used in this experiment. These were lettuce (*Lactuca sativa L*) and amaranth (*Amaranthus hybridus*). The seeds were purchased in Ouagadougou from NAKOSEM, a store specializing in the sale of market garden seeds. The selection criteria for the market garden produce were based on their availability throughout the year, and their appreciation by the local population. From a health point of view, this will enable us to assess the health risks associated with the consumption of these species, in order to raise farmers' awareness of these agricultural practices, as well as their capacity to accumulate heavy metals, as demonstrated by numerous previous studies [23].

2.1. Lettuce

This plant belongs to the Asteraceae family, genus Lactuca; species *Lactuca sativa*. It is an annual herbaceous plant, cultivated for its tender leaves eaten as a vegetable, generally raw and in salads. Lettuce requires soil rich in organic matter; however, it can be grown in any type of soil, such as sandy-loam, humus and muddy soils...etc. Soil pH should be between 6.0 and 8.0, as lettuce does not tolerate acidic soils (pH< 5.5) [26]. Lettuce is a cool summer vegetable; high temperatures and extreme light levels cause tissue hardening and rapid flowering. In good conditionslettuce germinates normally between 4 and 26°C over a period of 1-4 days, and grows best between 15 and 18°C (with a minimum of 7°C and a maximum of 24°C). When the temperature rises above 18°C, the plant moves from the head stage to the flowering stage [26]. When grown on contaminated soil, lettuce can accumulate the TMEs contained therein. Among the many vegetable species that accumulate heavy metals is lettuce. It has been considered a risk species cadmium accumulation [17].

2.2. Amaranth

Amaranth (*Amaranthus hybridus*) is an annual plant of the Amarantaceae family, belonging to the genus Amaranthus, cultivated for its edible leaves. It is a C4 plant, which means it is very demanding in terms of light energy. According to MADRPM [27], amaranth growth stops completely at 8°C. However, it grows well when daytime temperatures are above 25°C and night-time temperatures do not fall below 15°C. Amaranth is well suited to subsistence agriculture, thanks to its tolerance of adverse environmental conditions such as salinity, high acidity, drought and soil diversity [28]. However, a soil rich in organic matter is ideal for rapid, high yields, as seeds need good fertile soil and adequate moisture to germinate. Grubben [29] adds that soil with a pH > 6 is ideal for rapid growth and high yield.

3. Methods

3.1. Soil sampling

The soil used as an absolute control is a *Lixisol* in the village of Gampéla (12°25'N and 1°21'W) with known characteristics, having been used in numerous previous and recent studies in Burkina. It is located in the Saaba commune25 km east of Ouagadougou. The surface horizon (0 to 14 cm) was sampled. Ten (10) 10 kg pots of soil sieved to 2 mm were used for the experiment. The characteristics of the soil used are given in Table 1.

Table no.1 . Characteristics of the soil used as an absolute control

Soil variables	Values
Clay (%)	14,5
Silt (%)	28,05
Sand (%)	38,49
Texture	Sandy-silt
C (g kg^{-1})	0,73
N (g kg^{-1})	0,05
C/N	14
Total P (ppm)	1257
Assimilable P (ppm)	1,3
Calcium (Cmoles $^{(+)}$ Kg^{-1} sol)	1,81
Magnesium (Cmoles $^{(+)}$ Kg^{-1} sol)	1,85
Potassium (Cmoles $^{(+)}$ Kg^{-1} sol)	0,11
Sodium (Cmoles $^{(+)}$ Kg^{-1} sol)	0,04
S (Cmoles $^{(+)}$ Kg^{-1} sol)	3,81
CEC (Cmoles $^{(+)}$ Kg^{-1} soil)	7,28
Saturation rate	52
pH water	5,46

Soil samples were taken from two landfill sites. These were the Toudouwéogo landfill (TT) and the Saaba landfill (TS), taking into account their large surface area and use by gardeners. From each landfill, ten (10) 10kg pots of raw soil free of coarse waste were taken. The various soil samples were transported to IRD for preparation and use in the vegetation vase test.

3.2 Physical characterization of landfill waste

The characterization of solid waste in a landfill provides information on the quantity and composition of the various wastes in the landfill. Waste composition fluctuates geographically and over time. Thus, the results of a characterization campaign remain valid locally, for a given period, and enable us to choose the treatment method to be applied to the landfill.

Physical characterization followed three important stages: sampling of landfill soils, categorization of waste by sorting and determination of each waste fraction.

The first stage consisted of sampling five (05) kg of raw waste from each landfill site in ten (10) refuse heaps. After collection, the composite sample was transported to the IRD for sorting by manual extraction of coarse particles into categories and sub-categories according to waste type. Each waste category was then weighed to determine its total quantity in the sample.

3.3. Soil sample preparation

The soil fraction was sieved on a 2 mm diameter sieve to obtain the fine soil. The rejects and fine soil were weighed on a balance. Once the rejects had been removed, the fine soil was dried at room temperature and stored in two sterile plastic bags. One bag was sent to BUMIGEB for heavy metal analysis, while the other was used for other chemical analyses.

3.4. Experimental treatment device

The trial was conducted using the Fisher experimental design. The two (02) landfill soils were treated with lettuce or amaranth, giving the following treatments

- Toudouwéogo landfill soil under amaranth cultivation (TTA) ;
- Saaba landfill soil under amaranth (TSA) cultivation;

- Absolute control under amaranth cultivation (TAA) ;
- Toudouwéogo landfill soil under lettuce cultivation (TTL) ;
- Saaba landfill soil under lettuce cultivation (TSL) ;
- Absolute control under lettuce (TAL).

The set-up consisted of six (06) treatments with five (05) replicates. A total of thirty (30) pots were used during the experiment, fifteen (15) for amaranth and fifteen (15) for lettuce.

3.5. Conducting the test

The trial was conducted from April 05 to June 10, 2023. Ten (10) pots were filled with 10kg of Toudouwéogo landfill soil, ten (10) pots with 10kg of Saaba landfill soil and ten (10) pots with 10kg of control soil from Gampéla. A total of thirty (30) pots were used to sample soil from the Toudouwéogo and Saaba landfills, as well as control soil. Plastic pots with a volume of 10 liters were used. Moisture at field capacity was determined before sowing the seeds in each pot.

Field Capacity Moisture and Irrigation: In order to determine the quantity of water for pre-irrigation, the field capacity moisture (FCM) was determined on April 03, 2023 on three (03) samples of each treatment. By definition, the HCC of a soil corresponds to the water retained by the soil after a rain and a two- to three-day re-drying period, with the soil protected against evaporation. It was determined using the formula for moisture at mass field capacity in percent. The bottom of the pot is perforated and closed with a nylon screen with a mesh size of less than 2 mm. The dry weight (Ps) was determined. Next, the soil was saturated with water and the pot of soil was allowed to dry for 24 hours. The wet weight Ph is then determined. Applying the formula HCC = (Ph - Ps) / Ps, the moisture at field capacity was 0.150 kg (100 ml) of water per kg of control soil and 0.200 kg (200 ml) of water per kg of landfill soil. The HCC result showed that 1l was required for pre-irrigation of the control soil and 2l for the landfill soils. Pre-irrigation was carried out the day before sowing and irrigation twice (02) a week (Table n°2).

Table 2: Quantity of water used for each irrigation operation

Irrigation	Water quantity in L	Dates

	Control soil	Landfill soil		
HCC	1	2	03/04/2023	
Pre-irrigation	1	2	05/04/2023	
03 JAS - 22 JAS	0,5	0,5	04/09 04/28/2023	to
24 JAS - 46 JAS	0,5	1	04/30 05/22/2023	to
48 JAS - 60 JAS	1	2	05/24 06/10/2023	to

Legend: *HCC= Field Capacity Moisture; DAS= Day after sowing.*

Sowing: After determining the humidity at field capacity (HCC), which made it possible to obtain the watering dose, the seeds were sown precisely on April 06, 2023 in the various pots with 20 seeds for amaranth and 10 seeds for lettuce. The average germination rate in six (06) days for amaranth was 100% and 80% for lettuce. The plants were de-mulched one week after germination to retain two plants per pot.

3.6. Agronomic data collection

Observations and measurements were made on growth indicators, i.e. morphological and agronomic parameters of amaranth and lettuce. Data were collected at one-week intervals from the eighth (8th JAS) to the sixtieth (60th JAS). The parameters measured were :

- **Emergence rate (TXL):** This is the percentage of lettuce and amaranth plants that have emerged per pot. The emergence rate was determined for each plant on 7 $^{(-th)}$ DAS.
- **Plant height (HP):** Plant HP in cm was measured from day 8 to day 60 for amaranth and from day 49 to day 60 for lettuce. These measurements were taken using a ruler and a metric tape graduated in cm. They were measured from the crown to the terminal bud.
- **Diameter at crown (DC):** The DC in mm was measured with a caliper from 15th JAS to $60^{(th)}$ JAS for amaranth and from 35th JAS to 60th JAS for lettuce.

- **Number of leaves (NF), flowering plants (NFP) and flowers per plant (NFP):** NF, NFP and NFP were determined by counting leaves, flowering plants and flowers per plant, respectively, up to harvest.
- **Plant biomass:** The assessment of biomass production at harvest consisted of stripping, separating plant aerial and root biomass by treatment. Fresh above-ground biomass (AFB) and fresh root biomass (FRB) were determined by crop and landfill soil; total fresh biomass (TFB) was determined by weighing. These biomasses were then oven-dried at 40°C to a constant weight and weighed. Dry aerial biomass (BAS), dry root biomass (BRS) and dry total biomass (BTS) in g were also measured.

3.7. Processing plant and soil samples

a) Preparation of samples for chemical analysis

An aliquot of each composite soil sample was taken before sowing, in order to assess total Carbon (C), Nitrogen (N), Phosphorus (P) and Potassium (K) content, P_assimilable, K_dispisponible, pH H_2O and KCl and the current level soil TME contamination. In order to take into account the heterogeneity of the environment, a composite sample was created from the fine soil of the elemental samples considered during the landfill characterization. Two (02) samples (fine soil from the Toudouwéogo and Saaba landfills) were used for these analyses. These analyses were carried out at the CID INGENIERIE laboratory for the determination of mineral elements

When the plants were harvested, 75 days after sowing, soil samples were taken from the rhizosphere zone and mixed to obtain a composite sample according to the different treatments. The soil was sieved through a 2 mm diameter stainless steel sieve. A microbial respiration test was performed. This consisted in quantifying the release of CO_2 in the soil after incubation. pH H_2O and KCl were also determined. $CO_{(2)}$ and pH measurements were repeated 3 times for each composite soil sample per treatment.

For the analysis of heavy metals in plants, a composite sample of the dry biomass (aerial and root) after harvesting was used. For the determination of heavy metals, a composite sample of rhizospheric soils treated with landfill soil was taken. The fine soil (sieved to 2 mm) from this composite sample was used for the analyses. A total of 8 plant and soil samples (4 from the Toudouwéogo landfill and 4 from the Saaba landfill) were analyzed. Heavy metal contents were determined in the soil and in the plant organs of lettuce and amaranth at the Geochemistry Laboratory of the Bureau des Mines et de la Géologie du Burkina (BUMIGEB).

b) Determination organic, chemical and biological soil parameters and heavy metals

- **Determination of chemical and biological parameters**

The analysis of soil parameters consisted in determining :

- The pH H_20 according to the AFNOR method [30] of soils before sowing and at harvest. Acidity was measured in a soil suspension with a ratio of soil to distilled water or KCl (1N) equal to 1/2.5, using a pH meter with a combined electrode.
- Organic and total carbon were measured using the Walkley-Black method [31]. The organic matter content was obtained by multiplying the carbon content value by a coefficient equal to 1.724.
- Total nitrogen (Nt) by colorimetry after digestion using the Kjeldahl method [32].
- assimilable and total phosphorus was that of Bray [33].
- Total and available potassium, after extraction with a mixed solution of 0.1 N hydrochloric acid HCl and sulfuric acid H_2SO4, measured using the flame photometer after mineralization.
- Soil respiration was measured through the release of CO_2 from soil samples incubated by the Dommergues method [34] in the UMR Eco&sols laboratory at IRD Ouagadougou.

- **Determination of heavy metals and calculation of the soil pollution index**

Heavy metal assays were carried out in the laboratory of the Bureau des Mines et de la Géologie du Burkina (BUMIGEB) on soils before sowing and at harvest. Soil and plant samples were first dried in the shade at room temperature, then crushed and sieved through a 63-micrometer sieve. To prepare the liquid solution, test tubes were mixed with one (01) gram of each soil and plant sample, to which ten (10) ml of 70% concentrated nitric acid (HNO_3) was added. A microwave digestion system (MARS 6) was used to digest the reaction mixtures at a precise temperature. Digested samples were transferred to 100 ml volumetric flasks and made up to the mark with distilled water to ensure thorough homogenization. This dilution is compatible with the detection threshold of the analytical instrument. Once decanted, the test tubes were used to take each sample; the content of each TME was determined by flame Atomic Absorption Spectrometer (AAS). The instrument used was a Perkin Helmer AAnalyst100.

The soil pollution index (PI) is calculated as the ratio of metal concentrations in the soil based on limit values corresponding to tolerable levels of metal concentrations (ETM) in the soil according to AFNOR standard U44-41 [35].

$$\mathbf{I.P = (Cu/100 + Pb/100 + Zn/300) / 3}$$

The PI is a criterion for assessing the overall toxicity of a contaminated soil. A pollution index greater than 1 means that the soil is considered contaminated and toxic [19]

3.8. Statistical analysis

EXCEL 2021 was used for data entry. The Shapiro-Wilk test was used to test the normality of the agronomic data collected. Graphs were designed to analyze the evolution of morphological and physiological parameters of amaranth and lettuce, as well as soil respiration under the different treatments as a function of time. For each of the parameters evaluated, an analysis of variance (ANOVA) and a Tukey HSD test of separation of means at the 5% threshold were performed on morphological, physiological and biomass production parameters at harvest as a function of treatments. XLSTAT 4.1, 2023 (ADDINSOFT, 2023) was used for these analyses.

III. RESULTS AND DISCUSSION

1. Results

1.1 Soil characteristics of the Toudouwéogo and Saaba landfills

a) Physical characteristics

The results in Table 3 show that landfill waste has a heterogeneous composition, with a fine soil content of 35% for the Toudouwéogo landfill (TT) and for the Saaba landfill (TS). Unwanted refuse such as plastics, batteries, biomedical waste, glass and metals are also present. In the soil of the Toudouwéogo landfill, solid medical waste represents a rate of 2%, compared with 1.76% in the soil of the Saaba landfill.

Type of waste	Soil at the Toudouwéogo landfill (TT)		Saaba landfill floor (TS)	
	Weight (g)	%	Weight (g)	%
Putrescibles	457,5	9,20	445,7	8,91
Paper	202,2	4,04	41,50	0,83
Metals	122,70	2,50	87,30	1,75
Glass	377,2	7,54	321,30	6,43
Plastic	650,7	13,01	700,90	14,02
Cardboard	156,8	3,14	92,60	1,85
Textiles	185	3,70	81,00	1,62
Sanitary textiles	104,1	2,08	82,80	1,66
Composites	259,6	5,20	119,10	2,38
Solid medical waste	123,23	2,46	87,8	1,76
Fuels	592,8	10,9	946,80	18,94
Fine	1750,17	35,10	1993,20	39,86
Total quantity withdrawn	5000	100	5000	100

Table n°3. Physical characteristics of landfill soils

Legend: *TS=Sol initial de la décharge de Saaba ; TT=Sol initial de la décharge Toudouwéogo*

b) Chemical and mineral characteristics

Table 4 gives a chemical characterization of the landfill soils through parameters such as pre-planting pH H_20 and KCl, organic matter (OM), carbon (C), nitrogen (N), total phosphorus (P), potassium (K), available P-assimilable and Potassium. The average pH H_20 values recorded were neutral in the two pre-planting soils at Toudouwéogo (7.01) and Saaba (6.92). The average pH KCl values recorded for the Toudouwéogo (7.06) and Saaba (7.07) soils prior to sowing were higher than pH H_20.

C, N, P and K content (Table 4) varied between the two landfills. The MO content was 1.for TT (Toudouwéogo landfill soil) versus 1.for TS (Saaba landfill soil). The landfill soils had C/N ratios of 12 for TT and 10 for TS. Total soil P content was 5355.94 mg/kg and 5213.30 mg/kg for the Toudouwéogo and Saaba landfill soils respectively. Total potassium K was 778.20 mg/kg for TT versus 1119.84 mg/kg for TS.

Features Chemicals	**Landfill floor of Toudouwéogo (TT)**	**Landfill floor by Saab (TS)**
pH H_20	7,01	6,92
pH KCl	7,06	7,07
Total OM (%)	1,82	1,69
Total C (%)	1,06	0,98
Total N (%)	0,09	0,10
C/N	12	10
Total P (mg/kg)	5355,94	5213,30
Total K (mg/kg)	778,20	1119,84
P_assimilable (mg/kg)	3,27	2,32
K_dispisponible (mg/kg)	142,53	125,02

Table 4. Chemical characteristics of landfill soil

Legend: *MO = Organic matter; C = Carbon; N = Nitrogen; P = Phosphorus; K= Potassium; TS = Saaba landfill initial soil; TT = Toudouwéogo landfill initial soil.*

c) Heavy metal content

The results of the TME content in the landfill soils are presented in Table 5; they reveal a difference in the content of each TME in the two landfill soils. Zn is the element with the highest content in both soils, compared with the other elements. Both landfill soils have a pollution index well above 1 (8.15 for Saaba landfill soil and 7.50 for Toudouwéogo landfill soil). Compared with the AFNOR standard, the levels of the three ETMs (Cu, Pb and Zn) are well above the maximum levels for ETMs in agricultural soils.

Table 5. TME content in landfill soils before semi-filling and pollution index

Treatments	**Content (mg/kg**

		Cu	Pb	Zn	Pollution index (IP)
TS		112,40	131,12	6600,00	8,15
TT		171,40	162,36	5750,00	7,50
AFNOR standard	U44-41	100	100	300	1

Legend: *TS = Initial soil of the Saaba landfill; TT=Initial soil of the landfill*

1.2 Effect of landfill soils on morphological and agronomic parameters of amaranth and lettuce

a) Evolution of amaranth morphological parameters

- **Plant height**

The evolution of average plant heights according to the different treatments is shown in figure 2a. Analysis of the figure shows an increasing trend in plant height for the TTA treatment (Toudouwéogo under amaranth cultivation), followed respectively by the TSA (Saaba under amaranth cultivation) and TAA (Absolute Control under amaranth cultivation) treatments from 8th to 60th JAS. Table 6 shows a better development in plant height (HP) of amaranth ranging from 81.95± 6.88 cm under TTA versus 3.69± 6.88 cm under TAA. The ANOVA and TukeyHSD test of separation of means at the threshold of (5%) showed a highly significant difference (P< 0.0001) between plant heights under the TTA and TSA treatments compared with the TAA treatment (Table n°6).

- **Neck diameter**

Figure 2b shows that the TTA treatment improved the diameter growth of amaranth plants, followed respectively by the TSA and TAA treatments from 15th to 60th JAS. The observation in Table 6 shows a better growth in diameter at the amaranth crown 16.86± 0.88 mm under TTA versus 1.97±0.88 mm under TAA, which showed the weakest evolution in plant thickness. The ANOVA and TukeyHSD test of separation of means at the 5% threshold showed a highly significant difference (P< 0.0001) between the diameter at the crown under the TTA treatment compared with the TSA and TAA treatments (Table n°6)

- **Number of leaves and branches**

Figures 2c and 5d show that the TTA treatment increased the number of leaves and branching of amaranth plants, followed respectively by TSA and TAA from 8th to 60th JAS. Observation of Table 6 shows that the TTA treatment recorded the highest number of leaves on amaranth plants, i.e. 61.40± 7.07, compared with 8.40± 7.07 under TAA, which recorded the lowest number of plant leaves. As for the number of branches (NR), the TTA treatment recorded 14.00± 2.24 versus 0.00± 00 under TAA. The ANOVA and TukeyHSD test of separation of means at the 5% threshold showed a highly significant difference ($P < 0.0001$) in the number of plant leaves under the TTL treatment compared with the TAL and TSL treatments (Table 6)

- **Number of flowers and plants in bloom**

Figures 2e and 5f show that the TTA treatment increased the number of flowering and blooming amaranth plants, followed respectively by the TSA and TAA treatments from 49th to 60th days before harvest. Table 6 shows that the TTA treatment recorded an average of 2.20± 0.69 amaranth plants having flowered (NPF), compared with 0.00± 0.69 under TAA. For the number of flowers per plant (NFP), the TTA treatment recorded 2.20± 2.24 compared with 0.00± 00 under the TAA treatment.

- **Lift rate**

The emergence rate of amaranth (TXL) was determined on 7thJAS. Table 6 shows a very good emergence rate for all treatments (99± 1.20 for TAA; 98± 1.20 for TSA and 96 ±1.34 for TTA). The ANOVA and the TukeyHSD test of separation of means at the 5% threshold show that there is no significant difference ($P= 0.282$) between treatments in terms of emergence rate (table n°6).

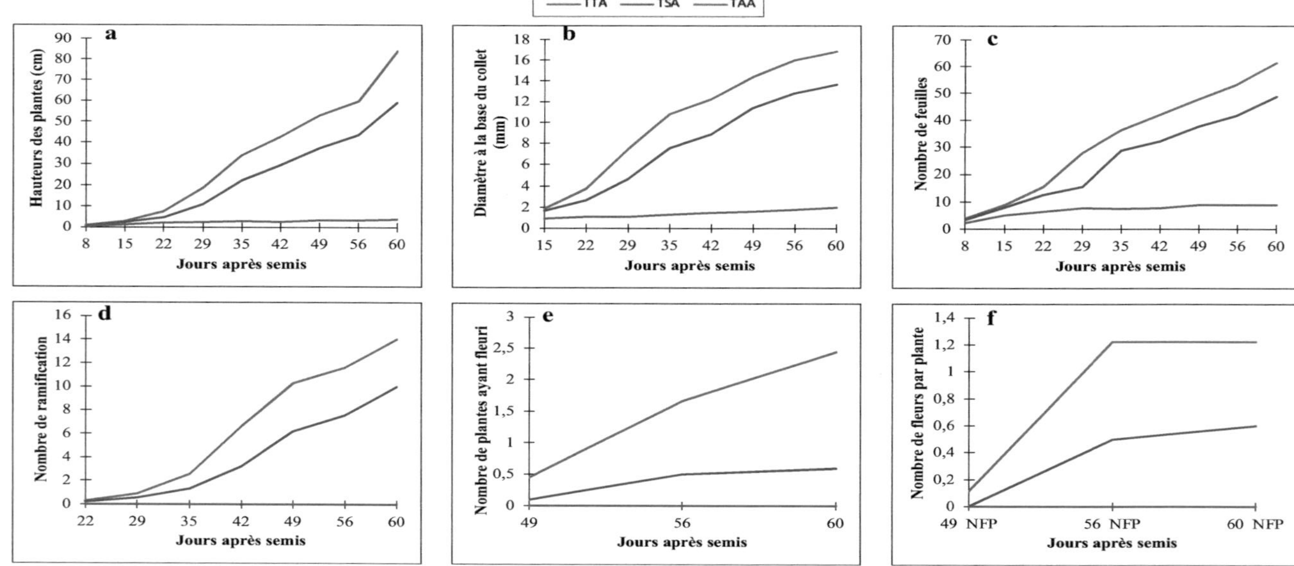

Figure no.2 : Evolution of morphological parameters of amaranth as a function of landfill soils

Legend: *TTA = Soil from Toudouwéogo dumps under amaranth cultivation; TSA = Soil from Saaba dumps under amaranth cultivation; TAA = Absolute control under amaranth cultivation.*

Treatments	TXL	HP	DP	NF	NR	NPF	NFP
	(%)	**(cm)**	**(mm)**				
TTA	96±1,34a	81,95±6,88b	16,86±0,88c	61,40±7,07b	14,00±2,24b	2,20±0,69a	2,20±0,69a
TSA	98±1,20a	58,78±7,26b	13,53±0,93b	50,44±7,45b	11,00±2,36b	0,67±0,73a	0,67±0,73a
TAA	99±1,20a	03,69±6,88a	01,97±0,88a	08,40±7,07a	00,00±2,24a	0,00±0,69a	0,00±0,69a
P	0,282	< 0,0001	< 0,0001	< 0,0001	0,000	0,088	0,088
Significance	NS	HS	HS	HS	HS	NS	NS

Table 6. Effect of landfill soils on amaranth morphological parameters

Legend: *TTA = Soil from Toudouwéogo dumps under amaranth cultivation; TSA = Soil from SAABA dumps under amaranth cultivation; TAA = Absolute control under amaranth cultivation; TXL= Seed emergence rate; HP= Plant height; DP= Plant diameter r; NF = Number of leaves; NR = Number of branches; NPF = Number of plants having flowered; NFP = Number of flowers per plant. Numbers are means ± standard errors of evaluated parameters. P: Probability according to ANOVA at 5% significance level. Means ± standard errors of the same column with the same letter do not differ significantly according to the TukeyHSD test at the 5% significance level.* $P < 0.05$*: significant (S);* $P \leq 0.01$*: very significant (TS);* $P \leq 0.001$ *(HS): highly significant;* $P \geq 0.05$*: not significant (NS)*

b) Evolution of lettuce morphological parameters

- **Plant height**

Figure 3a shows the evolution of average plant heights according to the different treatments. This figure shows that the TTL treatment improved lettuce plant height (HP), followed respectively by the TSL and TAL treatments from $49^{th}$to 60^{th}JAS. The observation in Table 7 shows improved lettuce plant height growth of 20.50± 0.84 cm under the TTL treatment, compared with 1.55± 0.79 and 1.50± 0.79 under the TSL and TAL treatments respectively. The ANOVA and TukeyHSD test of separation of means at the 5% threshold showed a highly significant difference (P< 0.0001) between plant heights under the TTL treatment compared with the TSL and TAL treatments (Table 7)

- **Neck diameter**

Figure 3b shows that the TTL treatment improved the diameter growth of lettuce plants, followed respectively by the TAL and TSL treatments from 35th to 60th JAS (Figure 3b). Table n°7 shows a better increase in the diameter at the neck (DC) of lettuce 9.32± 0.80 mm under TTL versus 3.16±0.76 mm under TSL, which showed the weakest evolution in plant thickness. The ANOVA and the TukeyHSD test of separation of means at the 5% threshold showed a highly significant difference (P< 0.0001) between DC under the TTL treatment compared with the TSL and TAL treatments (Table 7)

- **Number of leaves**

Figure 3c shows that the TTL treatment increased the number of leaves on lettuce plants, followed by the TAL and TSL treatments from 8th to 60th days. The average number of leaves (NF) varied from 02 to around 14 per plant. Table 7 shows that the TTL treatment recorded the highest number of lettuce plant leaves, 13.89± 0.87, compared with 6.40± 0.82 and 2.70± 0.83 respectively under TAL and TSL, which recorded the lowest numbers of plant leaves. The ANOVA and TukeyHSD test of separation of means at the 5% threshold showed a highly significant difference (P< 0.0001) between the number of plant leaves under the TTL treatment compared with the TAL and TSL treatments (Table 7)

- **Lift rate**

Table 7 shows a very good emergence rate (TXL) for all treatments (99±1.45 under TAL; 98±1.29 under TSL; 97±1.29 under TTL). The ANOVA and TukeyHSD test of separation of means at the threshold of (5%) shows that there is no significant difference (P= 0.499) between treatments in terms of emergence rate.

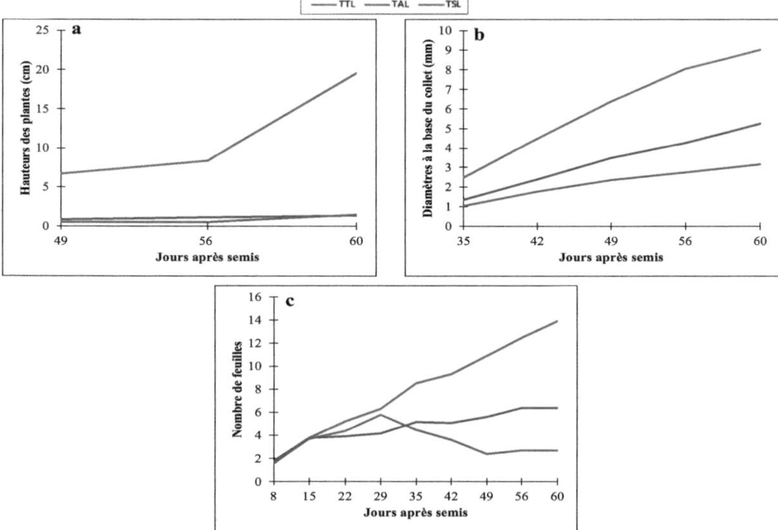

Figure no.3 : Changes in lettuce morphological parameters as a function of landfill soils

***Legend**: DAS = Days after sowing; TTL = Toudouwéogo landfill soil under lettuce cultivation; TSL = Saaba landfill soil under lettuce cultivation; TAL = Absolute control under lettuce cultivation.*

Table 7. Effect of landfill soils on lettuce morphological parameters

Treatments	TXL	HP	DP	NF
	(%)	(cm)	(mm)	
TTL	97±1,29 a	20,50±0,84b	9,32±0,80b	13,89±0,87c
TSL	98±1,29 a	01,55±0,79a	3,16±0,76a	02,70±0,83a
TAL	99±1.45 a	01,50±0,79a	5,27±0,76a	06,40±0,83b
P	0,499	<0,0001	<0,0001	<0,0001
Significance	TS	HS	HS	HS

Legend: *TTL = Soil from Toudouwéogo dumps under lettuce cultivation; TSL = Soil from Saaba dumps under lettuce cultivation; TAL = Absolute control under lettuce cultivation; TXL= Seed emergence rate; HP= Plant height; DP= Plant diameter r; NF = Number of leaves; Numbers are means ± standard errors of evaluated parameters. P: Probability according to ANOVA at 5% significance level. Means ± standard errors of the same column with the same letter do not differ significantly according to the TukeyHSD test at the 5% significance level. P < 0.05: significant (S); P ≤ 0.01: very significant (TS); P ≤ 0.001 (HS): highly significant; P ≥ 0.05: not significant (NS)*

c) Effect of landfill soils on biomass production of amaranth and lettuce

- **Fresh aerial and root biomass**

Fresh above-ground biomass (BAF) of amaranth under the TTA treatment varied significantly (P <0.0001) from 622.92± 36.91 to 404.98± 36.91 g under the TSA treatment and 0.70±41.27 under TAA (Table n°8). BAF values under the TSA and TTA treatments were significantly different. Fresh root biomass (FRB) did not vary significantly between TTA (51.54±6.48 g) and TSA (46.92± 6.48) g, but was highly significant (P <0.0001) g under TAA (0.04±7.25).

For lettuce, AFB under the TTL treatment varied significantly (P <0.000) from 155.82±12.93 g to 11.84± 12.93 g under TSL and 8.63± 14.45 g under TAL. The ANOVA and TukeyHSD test of separation of means at the (5%) threshold show that there is no significant difference between the TSL and TAA treatments. Fresh root biomass (RCB) was highly significant (P< 0.0001) under TTL (4.43± 0.54 g) compared with that obtained under TSL (0.42± 0.54g) and TAA (0.47± 0.60 g).

- **Total fresh biomass**

The total fresh biomass (BTF) of both landfill soils for crops was significantly increased compared with the absolute control soil. Amaranth under the TTA treatment significantly ($P < 0.0001$) varied from 674.46± 41.05 g to 451.90± 41.05 g under TSA and 0.74± 45.89 g under TAA . For lettuce BTF, the TTA treatment significantly ($P < 0.0001$) increased from 160.25± 13.46 g to 12.26± 13.46 g under TSA and 9.09± 15.04 g under TAA (Table 8).

- **Dry aerial and root biomass**

Dry above-ground biomass (DAB) of amaranth varied significantly from 81.5± 4.49 under TTA to 51.3± 4.49 g under TSA treatment and 0.1± 5.01g under .
(Table 8). Dry root biomass (DRB) of amaranth did not vary significantly between the TTA (7.04± 1.30 g) and TSA (7.36± 1.30 g) treatments, but was significantly improved ($P < 0.0001$) compared with that under TAA (0.02± 1.45 g).

For lettuce, BAS varied significantly under TTL (14.28± 1.18g compared with both treatments TSL (1.61± 1.18g) and TAL (0.76± 1.32g). BRS increased significantly ($P < 0.0001$) under TTL (0.94± 1.11g) compared with TSL (0.10± 0.11g) and TAA (0.11± 0.12g).

- **Total dry biomass**

The total dry biomass (TDB) of both landfill soils for crops was significantly improved compared to that recorded under the absolute control. The TTA treatment significantly increased ($P < 0.0001$) by 88.50± 5.22g compared with 58.66± 5.22g under TSA and 0.17± 5.84g under TAA. TTA treatment significantly increased ($P < 0.0001$) the BTS of lettuce, from 15.22± 1.27g to 1.71± 1.27g under TSA and 0.87± 1.42g under TAA (Table n°8).

Table 8. Effect of landfill soils on biomass production of amaranth and lettuce at harvest

Treatments	**BAF (g)**	**RCW (g)**	**BTF (g)**	**LOW (g)**	**BRS (g)**	**BTS (g)**
TTA	622,92±36,91c	51,54±6,48b	674,46±41,05c	81,5±4,49c	7,040±1,30b	88,50±5,22c
TSA	404,98±36,91b	46,92±6,48b	451,90±41,05b	51,3±4,49b	7,36±1,30b	58,66±5,22b
TAA	0,70±41,27a	0,04±7,25a	0,74±45,89a	0,15±5,01a	0,02±1,45a	0,17±5,84a
P Significance	<0,0001 HS	0,0001 HS	<0,0001 HS	<0,0001 HS	0,005 HS	<0,0001 HS
TTL	155,82±12,93b	4,43±0,54 b	160,25±13,46b	14,28±1,18b	0,94±0,11b	15,22±1,27b
TSL	11,84±12,93a	0,42±0,54 a	12,26±13,46a	1,61±1,18a	0,10±0,11a	1,71±1,27a
TAL	8,63±14,45a	0,47±0,60 a	9,09±15,04a	0,76±1,32a	0,11±0,12a	0,87±1,42a
P Significance	<0,0001 HS	0,000 HS	<0,0001 HS	<0,0001 HS	0,000 HS	<0,0001 HS

Legend: *TTA = Soil from Toudouwéogo dumps under amaranth cultivation; TSA = Soil from Saaba dumps under amaranth cultivation; TAA = Absolute control under amaranth cultivation; TTL = Soil from Toudouwéogo dumps under lettuce cultivation; TSL = Soil from Saaba dumps under lettuce cultivation ; TAL = Absolute control under lettuce cultivation; BAF = Fresh above-ground biomass; BRF = Fresh root biomass; BAS = Fresh above-ground biomass; BRS = Fresh root biomass; BTF = Total fresh biomass; BTS = Total dry biomass; Numbers are means ± standard errors of evaluated parameters. P: Probability according to ANOVA at 5% significance level. Means ± standard errors of the same column with the same letter do not differ significantly according to the TukeyHSD test at the 5% significance level. P < 0.05: significant (S); P ≤ 0.01: very significant (TS); P ≤ 0.001 (HS): highly significant; P ≥ 0.05: not significant (NS).*

1.3. Soil fertility of landfill sites where amaranth and lettuce are grown

a) Effect on soil respiration

Figures 4a and 4b show the evolution of CO_2 release by soil microbial biomass for amaranth and lettuce respectively, as a function of periods ranging from one hour after incubation (1HAI) to 22 days after incubation (22JAI). These figures show that the curves formed by the TTA, TSA, TTL, TSL treatments as a function of incubation periods are well above the curves for the TAA and TAL treatments.

Table 9 shows that soil microbial $CO_{(2)}$ release under the TTA and TSA treatments is significantly higher under amaranth cultivation compared to the TAA control at 96HAI, 5JAI, 11 JAI,16 JAI and 22 JAI. ANOVA at the 5% threshold shows a highly significant difference in CO_2 release ($p< 0.001$) between the two treatments TTA; TSA versus TAA.

Under lettuce, the TTL and TSL treatments significantly increased microbial $CO_{(2)}$ release from soils at 1 HAI, 96HAI, 16 JAI and 22 JAI compared with the TAL control (Table 9). At 5HAI and 11HAI, the 3 treatments affected microbial $CO_{(2)}$ release differently, with the highest values obtained under the TSL treatment.

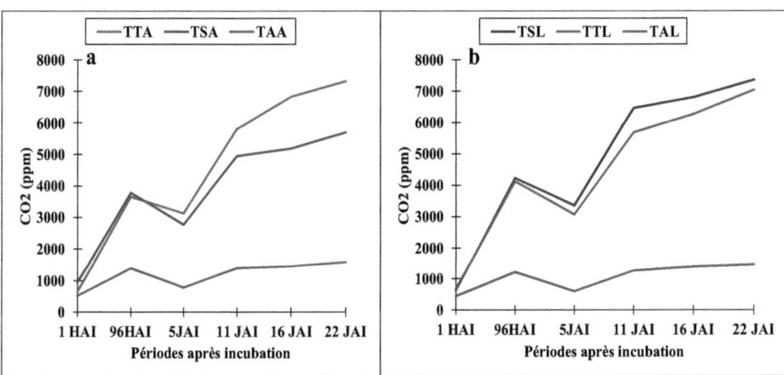

Figure n°4 : Evolution of $CO_{(2)}$ release from landfill soils under amaranth (a) and lettuce (b) cultivation as a function of incubation periods

Legend: *TTA = Soil from Toudouwéogo dumps under amaranth cultivation; TSA = Soil from Saaba dumps under amaranth cultivation; TAA = Absolute control under amaranth cultivation; TTL = Soil from Toudouwéogo dumps under lettuce cultivation; TSL = Soil from Saaba dumps under lettuce cultivation;*

TAL = Absolute control under lettuce cultivation; HAI = Hours after incubation; JAI= Days after incubation.

Table 9. Quantity of CO_2 released from soils under amaranth and lettuce cultivation at each incubation period

Treatments	1 HAI	96HAI	5JAI	11 JAI	16 JAI	22 JAI
	CO_2 (ppm)					
TTA	645.00±85.03ab	3655,00±246,69 b	3135,00±149,81b	5825,00±611,76b	6845,00±584,04b	7330,00±499,49b
TSA	956,67±69,43b	3783,33±201,42b	2780,00±122,32b	4960,00±499,50b	5180,00±476,86b	5683,33±407,83b
TAA	530,00±69,43a	1396,67±201,42a	780,00±122,32a	1393,33±499,50a	1450,00±476,86a	1583,33±407,83a
P	0,018	0,001	<0,0001	0,004	0,002	0,001
Significance	TS	HS	HS	HS	HS	
TSL	636,67±31,19b	4236,67±380,79b	3346,67±47,52c	6470,00±179,42 c	6796,67±149,35 b	7035,00±197,07b
TTL	700,00±25,47b	4130,00±310,92b	3065,00±38,80b	5675,00±146,49b	6265,00±182,91 b	7366,67±241,36b
TAL	463,33±25,47a	1230,00±310,92a	0603,33±38,80a	1266,67±146,46a	1386,67±149,35a	1453,33±197,07a
P	0,004	0,002	<0,0001	<0,0001	<0,0001	<0,0001
Significance	HS	HS	HS	HS	HS	

Legend: *TTA = Soil from Toudouwéogo dumps under amaranth cultivation; TSA = Soil from Saaba dumps under amaranth cultivation; TAA = Absolute control under amaranth cultivation; TTL = Soil from Toudouwéogo dumps under lettuce cultivation; TSL = Soil from Saaba dumps under lettuce culture; TAL = Absolute control under lettuce culture; HAI = Hours after incubation; JAI= Days after incubation; Numbers are means ± standard errors of evaluated parameters. P: Probability according to ANOVA at 5% significance level. Means ± standard errors of the same column with the same letter do not differ significantly according to the TukeyHSD test at the 5% significance level.* $P < 0.05$*: significant (S);* $P \leq 0.01$*: very significant (TS);* $P \leq 0.001$ *(HS): highly significant;* $P \geq 0.05$*: not significant (NS).*

b) Effects on soil hydrogen potential

Table 10 shows average pH H_2O and pH KCl values at harvest. The mean pH H_2O values for the Toudouwéogo soil treatments (7.59±0.02 under amaranth 7.75±0.01 under lettuce) and Saaba soil (7.64±0.02 under amaranth to 7.76±0.01 under lettuce) were highly significant compared with the absolute control treatment (5.94±0.01 under amaranth 6.45±0.01 under lettuce). No significant difference was recorded for pH H_2O between the TTA and TSA treatments on the one hand, and TTL and TSL on the other. However, a highly significant difference was observed between the landfill soil and the control soil ($P < 0.0001$).

The average pH KCl values of the different soils after harvesting were slightly below the pH water values for all crops (Table 10). Under amaranth, pH KCl was 7.34±0.01 under TTA, 7.51± 0.01 under TSA versus 5.43± 0.01 under TAA. Under lettuce, the pH treatment was 7.52± 0.01 under TTL, 7.54±0.01 under TSL and 5.98± 0.01 under TAL. A significant increase in pH KCl was observed between TTA, TSA and TSA for amaranth cultivation. The TTL and TSL treatments significantly raised pH KCl values compared with TAL.

Table 10. Comparison of the hydrogen potential of landfill soils under amaranth and lettuce cultivation

Treatments	pH H_2O	pH KCl
TTA	7,59±0,02b	7,34±0,01b
TSA	7,64±0,02b	7,51±0,01c
TAA	5,94±0,01a	5,43±0,01a
P	<0,0001	<0,0001
Significance	TS	HS
TTL	7,75±0,01b	7,52±0,01b
TSL	7,76±0,01b	7,54±0,01b
TAL	6,45±0,01a	5,98±0,01a
P	<0,0001	<0,0001
Significance	HS	HS

Legend: TTA = *Soil from Toudouwéogo dumps under amaranth cultivation; TSA = Soil from Saaba dumps under amaranth cultivation; TAA = Absolute control under amaranth cultivation; TTL = Soil from Toudouwéogo dumps under lettuce cultivation; TSL = Soil from Saaba dumps under lettuce cultivation; TAL = Absolute control under lettuce cultivation; Numbers are means ± standard errors of the parameters evaluated. P: Probability according to ANOVA at 5% significance level. Means ± standard errors of the same column with the same* *letter do not differ significantly according to the TukeyHSD test at the 5% significance level.* $P < 0.05$*: significant (S);* $P \leq 0.01$*: very significant (TS);* $P \leq 0.001$ *(HS): highly significant;* $P \geq 0.05$*: not significant (NS).*

1.4 Effect of amaranth and lettuce plants on the phytoremediation of landfill soil in terms of TMEs

The results of the TME content in the soil after cultivation are shown in Table 11. The two plant species cultivated on the soils of the Toudouwéogo and Saaba landfills accumulated quantities of metals in their vegetative organs and thus contributed to the reduction of their levels in the soil. In terms of proportion, there was a reduction (20% Cu; Pb and Zn) in the content of these three TMEs in the Saaba soil under lettuce cultivation (TSL); for (28% Cu; Pb and Zn) the same soil under amaranth cultivation (TSA). In the case of the Toudouwéogo landfill soil, a drop of (13% Cu; 23% Pb and 19% Zn) in soil metal content was observed after amaranth cultivation (TTA), compared with (10% Cu; Pb and Zn) under lettuce cultivation (TSA). Thus, in the same soil, there is no great difference in TME reduction between the two plant species. Compared with the CMR standard (10 mg/kg Cu; 3 mg/kg Pb; 0.5 mg/kg Zn), the two plant species grown on the soils of the two landfills accumulate levels well above this standard.

The pollution index of landfill soils under cultivation of plant species (Table n°12) was considerably reduced compared with the initial soils before cultivation (Table n°5 and n°12). Also for this index, a difference was observed between the soils of Saaba (IP=1.6) and Toudouwéogo (IP=6), far exceeding the pollution standard (IP≤ 1).

Table 11. TME content of amaranth and lettuce plants and landfill soil

Treatments	Plant and soil samples	Content (mg/kg)		
		Cu	**Pb**	**Zn**
			Plant	
TTL	Lettuce plant under the soil of the Toudouwéogo landfills	21,44	19,26	1011
TSL	Lettuce plant under the soil of Saaba's rubbish dumps	12,45	19,26	123,6
TTA	Amaranth plant under the soil of the Toudouwéogo landfill site	94,63	32,45	423
TSA	Amaranth plant under the soil of Saaba's rubbish dumps	14,9	25,85	136,7
Standard	CMR (Edible leaf vegetables)	≤ 10	≤ 3	≤ 0,5
			Soil	
TTL	Toudouwéogo landfill soil under lettuce cultivation	155,18	138,06	4590
TSL	Saaba landfill soil under lettuce cultivation	90,5	118,27	837
TTA	Toudouwéogo landfill soil under amaranth cultivation	148,18	124,18	4655
TSA	Saaba landfill soil under amaranth cultivation	80,58	118,27	906
Standards	AFNOR U44-41	≤ 100	≤ 100	≤ 300

Legend: *TTA = Soil from Toudouwéogo dumps under amaranth cultivation; TSA = Soil from Saaba dumps under amaranth cultivation; TAA = Absolute control under amaranth cultivation; TTL = Soil from Toudouwéogo dumps under lettuce cultivation; TSL = Soil from Saaba dumps under lettuce cultivation; CMR: maximum regulatory concentration*

Table 12. Pollution index (PI) of soils under amaranth and lettuce cultivation

Treatments	Soil samples	IP
TTL	Toudouwéogo landfill soil under lettuce cultivation	6,07
TSL	Saaba landfill soil under lettuce cultivation	1,62
TTA	Toudouwéogo landfill soil under amaranth cultivation	6,08
TSA	Saaba landfill soil under amaranth cultivation	1,66
	AFNOR U44-41 standard	**IP 1≤**

Legend: *TTA = Soil from Toudouwéogo dumps under amaranth cultivation; TSA = Soil from Saaba dumps under amaranth cultivation; TAA = Absolute control under amaranth cultivation; TTL = Soil from Toudouwéogo dumps under lettuce cultivation; TSL = Soil from Saaba dumps under lettuce cultivation; IP= Pollution index.*

2. Discussion

2.1. Soil characteristics of the Saaba and Toudouwéogo landfills

The various chemical and microbiological analyses carried out on landfill soils reveal that they contain organic matter and nutrients. Many authors have shown that organic amendments provide essential nutrients for plants and help maintain pH levels around neutral, making nutrients more available [36].

In our study, we obtained neutral pH H_20 values in landfill soils prior to sowing, which would have favorably increased the levels of available potassium and assimilable phosphorus in the soil samples. pH is an essential parameter for plants. Indeed, when pH is close to neutral (6.5-7.5), the availability of essential nutrients such as nitrogen, phosphorus, potassium and other trace elements in the soil is maximized [37]. The pH KCl values recorded remain slightly higher than the pH H_20 values of pre-planted soils. pH KCl is generally well below pH H_2O except in soils containing high levels of metal oxides [38]. This result already supports the hypothesis of a higher presence of metal cations in pre-planting landfill soil samples. Our results are similar to those of a study carried out on landfill soil in Ahfir-Saidia (Morocco), which revealed that landfill soils are considerably contaminated by numerous metals, with concentrations close to or well above the recommended threshold values [39]. However, pH KCl values were below pH H_2O in the post-harvest soils. Overall, these soil samples were quite low in metal oxides [38]. This could explained by the fact that plants contributed to the depletion of metal oxides in the soil. A large proportion of metals would therefore be sequestered in plant tissues [40].

The organic matter (OM) content of landfill soil samples shows a fairly high OM content. These results are in line with those of Niang, who states that municipal solid waste is a deposit of organic matter [21]. It contains very high levels of fresh and transient organic matter, a precursor to humus, and also contains mineral compounds in various forms, which can contribute to soil fertilization and thus increase crop yields [41]. OM content can be explained by the release of carbon from the decomposition of soil waste. A high carbon content induces a high organic matter content. pH also plays a major role in carbon content, since a rise in pH increases the mineralization rate of organic matter, and thus the soil's carbon, nitrogen and phosphorus content, which in turn promotes microbial activity. This organic matter improves the biological and physico-chemical properties of soils, and is a source of nutrients (Potassium-Total) [42].

The average C/N ratio of 12 in the Toudouwéogo soil and 10 in the Saaba soil reflects the level of mineralization of organic matter in the soils. The values of this ratio obtained in the soils correspond to those indicating good mineralization of organic matter, i.e. between 8 and 12 [43]. According to Salomon BOUDA [37], the C/N ratio is an index of organic matter quality. The higher the C/N ratio< 12, the greater the biological activity and mineralization.

Microorganisms are key players in the major processes of matter transformation and energy flow in the soil. As part of our study, in order to assess microbiological activity, we measured the release of CO_2 from harvested soils, which results from the degradation of carbonaceous substrates by microorganisms over a two-week period. The results obtained show a high intensity of biological activity by microorganisms in landfill soils compared with control soils. This is supported by Evariste et *al* [44], who confirm that the quantity, quality and state of decomposition of organic matter can have a considerable influence on CO_2 release. A study carried out on landfill sites in the city of Kisangani (DR Congo) by Joël M. [45] confirms that landfill sites are major $CO_{(2)}$ emitters. The results of this study show that the quantities of carbon released by landfills are potentially significant compared with other types of soil, and are positively correlated with moisture content and temperature. Ago E.E. *et al* [46], explain that the release of CO_2 in soil is linked to various biotic factors such as soil biodiversity, and above all microbial biomass and the type of population it comprises.

2.2 Effects of landfill soils on growth and biomass production of amaranth and lettuce

Landfill soils increased plant growth and biomass production over time. These performances could be explained by the fertilizing properties of landfill soils through the input of carbon (C/N= 12), nutrients (assimilable P 3.27g/kg; available K 142.53g/kg) and indirectly by the physico-chemical properties of the soil, notably pH (7.01), soil moisture and soil biological activity. Firstly, waste increases the quantities of available nutrients, such as N and P. Secondly, plant growth is enhanced by the increased availability of mineral elements. Several authors have demonstrated the positive effect of organic amendments on increasing agricultural yields [47] [48]. Indeed, good assimilation of nutrients such as nitrogen, phosphorus, potassium and magnesium by plants promotes their growth and development [49]. It should be noted that good plant growth and resistance to various stresses (e.g. parasitism) result not only from a good supply of N, P and K elements, but also from sufficient availability of secondary elements, trace elements and various activators [50]. According to the same author [50], humus favors the supply of trace elements to plants, and provides growth activators that act at very low doses. It is therefore easy to understand why plants grown on landfill soils grow better, as these soils are much richer in organic matter and are better able to cover their needs than plants grown on control soils. The positive effect of landfill soils on plant growth was mainly due to improvements in the physico-chemical and biological quality of the soil, the rate of nutrient diffusion and the water retention capacity of the soil. The more organic matter a soil contains, the more fertile and productive it becomes. Tessier [51] also reported that organic inputs stimulate biological activity by increasing soil humus through the decomposition of crop residues.

2.3 TME concentration in soils and accumulation in plants

a) ETM concentration in soils (IP)

Soils from the uncontrolled landfills studied have high levels of TMEs. This is probably due to the heterogeneity of the wastes that make up the landfills. Waste electrical and electronic equipment, tires, plastics and medical waste present in the dumps may be sources of the high concentration of heavy metals in the soil [52]. The results show that the mean values of trace metal concentrations in soils vary according to metal element and sampling site. These results are comparable to those of several authors [53] [54] [55] [56] who have shown that total trace metal element (TME) levels in soils vary according to metal element, sampling site and source of contamination.

Analysis of the soil from the two dumps prior to cultivation shows that Pb, Cu and Zn levels are well above the AFNOR U44-41 standard (Cu 100 mg /kg; Pb 100 mg/kg and Zn 300 mg/kg), as well as the worldwide average values for tolerable heavy metal levels in contaminated soils (30 mg/kg for copper; 35 mg/kg for lead and 90 mg/kg for zinc [57]. These results are confirmed by Soclo [58], who has shown that compost made from household waste collected directly from illegal dumps contains heavy metals in relatively high proportions. With an IP of over 1, landfill soils are highly contaminated [19].

Zinc is the metal element with the highest concentrations (maximum and average) in soils. High concentrations of Zn in soils, compared with other elements, have also been reported by numerous authors [53] [56]. The zinc content of 5750 mg/Kg for the Toudouwéogo soil and 6600 mg/Kg for the Saaba soil indicates that the values are well above the accepted standard for agricultural soil quality (200 mg/Kg) according to Canadian guidelines for agricultural soil quality. Several studies have shown that municipal waste contains significant quantities of metals in general and zinc in particular [23] [59].

According to Canadian guidelines for agricultural soil quality, the maximum permissible concentration of lead in soil is 70 mg/kg. However, the results of our study revealed that lead levels in the two landfill soils all exceeded the standard (131 mg/kg for Saaba soil and 162 mg/kg for Toudouwéogo soil). This presence of lead may be attributable to contamination by used batteries found in the waste following poor sorting at the base prior to landfilling. Similar results obtained in a study carried out in Morocco on the soils of the Ahfir-Saidia uncontrolled landfill contaminated by heavy metals [39], which showed that Pb was present throughout the landfill and that its distribution varied according to depth.

As for the rhizospheric soils of the two landfills after amaranth and lettuce cultivation, the PI results are still higher than 1, but still lower than the PI of the landfill soils before sowing. Hence the conclusion that soils are still polluted. Nonetheless, TME levels fell considerably with the application of these species in crops. Overall, Zn was greatly reduced by 85% in the Saaba landfill soil, and very slightly in the Toudouwéogo soil (20%). Cu and Pb were slightly reduced by 20% and 10% respectively in the Saaba landfill soil, compared with 10% and 23% respectively in the Toudouwéogo landfill soil

b) ETM accumulation in plants and phytoremediation

TME content in plants differed according to plant species, type of TME and level of soil contamination, as might be expected. Lettuce grown in landfill soils accumulated TMEs in its various organs. Heavy metal levels in the above-ground biomass of Saaba landfill soils ranged from 12mg/kg Cu, 19.26mg/kg Pb and 1011mg/kg Zn to 21.44mg/kg Cu, 19.26mg/kg Pb and 1011mg/kg Zn for Toudouwéogo landfill soil. TME levels were well above the limit values above which phytotoxicities are possible (20-200 mg/kg for Cu, 30-300 mg/kg for Pb, 100-400 mg/kg for Zn) [60]. Similar results were obtained by Pasquini [22], who analyzed metals in lettuce from soils amended with municipal waste ash in Nigeria. Concentrations ranged from 57.5 to 182.1 mg/kg for zinc, 5.5 to 13.9 mg/kg for Cu and 5.8 to 11.6 mg/kg for lead. Among the many vegetable species that accumulate TMEs is lettuce [17]. It has been considered a high-risk species in terms of TME accumulation in aerial organs. Work by Agbossou [4] has shown that lettuce's rapid growth accelerates the uptake of the nutrients it needs from the soil, notably zinc and copper. In addition, the results of ADEME [14] showed the great capacity of fast-growing species such as lettuce to accumulate zinc, cadmium and copper.

Amaranth growing on the two landfill soils accumulated TMEs at levels ranging from 14mg/kg Cu, 25mg/kg Pb and 136mg/kg Zn for the Saaba landfill to 94.6mg/kg Cu, 32.45mg/kg Pb and 423mg/kg Zn for the Toudouwéogo landfill. We note a mobility of TMEs from the soil to the biomass and intoxication by exceeding the maximum reference concentration (CMR) for the three metals as recommended. Zinc remains the most accumulated element. The high Zn concentrations recorded in amaranth and lettuce on both soils can be explained by the soil's high zinc content, its bioavailability and zinc's strong affinity for these vegetables. These results are in line with those of Ondo *et al* [61], who demonstrated that the amaranth plant can accumulate high Zn concentrations if the soil on which it is grown contains moderately higher levels of Zn

Overall, Cu accumulation was low in both species compared with Zn. The low Cu levels recorded in the organs of the 2 plant species may be explained by competition between Zn and Cu. Indeed, according to Kabata-Pendias [62], Cu and Zn are apparently absorbed by the same mechanism, and their combined presence in the soil may limit their uptake.

In terms of health, the results obtained on the MTE content of the two plant species enable us to assess the toxicity of metals, defined as the concentration of the metal in the edible parts of plants, not exceeding the tolerable limit in humans. Comparing these results with the maximum concentrations of metallic elements in foodstuffs intended for human consumption, as set by the FAO/WHO [63] (73 mg/kg for Cu, 100 mg/kg for Zn and 0.3mg/kg for Pb), our two plant species are well above the levels authorized for human consumption. Zinc is an essential element required by plants and humans in appropriate proportions, but when concentrations exceed the normal required range, this can cause disease leading to dysfunction and unwanted accumulation in various parts of the body.

RECOMMENDATIONS

Recommendations for municipal authorities, Ouagadougou residents who produce waste, and farmers to reduce TME levels in soils from uncontrolled landfills:

- Sort and decontaminate household waste before depositing it in any landfill to reduce TME contamination;
- Sensitize market gardeners to change their behavior so that unsorted or poorly sorted household waste is not used as compost;
- Composting the waste from these landfills could reduce the level of TMEs;
- Particular attention needs to be paid to landfill soils that are used as fertilizer in urban fields and gardens, as the uptake of TMEs by plants grown in the fields can lead to contamination of the trophic chain.

CONCLUSION

The aim of this work was to evaluate the effects of anarchic landfill soils on amaranth and lettuce production, and the phytoremediation of these two plants in heavy metals. Characterization of these landfill soils was necessary in order to determine their composition. The results of this characterization confirmed our first hypothesis, namely that the two (02) anarchic public dumps investigated contain organic substrates, mineral compounds and TMEs. The experimental study carried out with lettuce and amaranth on soils with high concentrations of ETMs, notably Cu, Pb and Zn, showed that these plants can establish themselves on soils with high concentrations of these ETMs. The results show that lettuce and amaranth accumulate ETMs differentially according to soil type and ETM. Amaranth exported more ETMs, notably Cu, Pb and Zn, than lettuce.

From an agronomic point of view, the two plant species grown on landfill soils promoted good vegetative growth and better biomass yields compared with the control soil. The landfill soils thus enabled a significant increase in plant height and biomass production in the vegetation vase. Our results confirm our initial hypotheses that soils derived from urban waste contaminated with heavy metals influence the growth and biomass production of cultivated plants.

The ETM pollution index (Cu, Pb and Zn) in rhizospheric landfill soils after the experiment was significantly reduced compared with landfill soils before sowing. Lettuce and amaranth helped to clean up these soils by extracting TMEs. These results confirm our hypothesis that plant species accumulate Cu, Pb and Zn from soils, thereby helping to clean up landfill soil. Nevertheless, the levels of these TMEs in these landfill soils after experimentation remain well above the tolerable thresholds for TMEs in contaminated soils. As a result, precautions must be taken when using these soils, as their uncontrolled inputs could lead to an accumulation of TMEs in the surface layers of the soil and, consequently, to contamination of agricultural produce. Continuous cultivation of these plants on landfill soils, while taking care to destroy the biomass recovered over a period of years, could lead to a considerable reduction in the ETM content of landfill soils. The choice of using these species in this study is interest from a health point of view, given that they are edible and widely consumed in Burkina Faso in general and in Ouagadougou in particular. The results obtained will serve as a database for raising farmers' awareness of the health risks associated with this farming practice .

OUTLOOK

To better capitalize on these research results, future research could :

- Repeat the experiment in the field to assess the content of these heavy metals in cultivated plants;
- Continuously cultivate these plants on landfill soils until an acceptable level of TMEs is reached in the soil, taking care to destroy the biomass recovered over the years. In this way, the soil can eventually be decontaminated by the continuous cultivation of certain plant species.

BIBLIOGRAPHICAL REFERENCES

1. **Agueh V, Degbey C.C, Sossa-Jerome C, Adomahou D, Paraiso M.N, Makoutode M, Fayomi B, 2015.** Level of contamination of market garden produce by toxic substances at the Houéyiho site in Benin. Houéyiho: Int. J. Biol. Chem. Sci. Vol. 9(1).

2. **Kitobo W, 2009.** Dépollution et valorisation des rejets miniers sulfurés du Katanga cas des tailling de l'ancien concentrateur de Kipushi. Katanga: doctoral thesis, Université de liège, faculté des sciences appliquées.

3. **Article R.2224-16, 2007** code générale des collectivités térritoiles;France: discharge of sewage sludge into the aquatic environment.

4. **Agbossou K. E, Sannyi M. S, Zokpodo B, Ahamide B and Guedegbe H. J,2009.** Evaluation qualitative de quelques légumes sur le périmètre maraiche de Houéyibo à cotonou au sud Bénin. Cotonou: Bulletin de la recherche Agronomique du Bénin.

5. **Nordberg M, Nordberg G, Jin T, 2004.** Health impacts of cadmium exposure and its prevention. BioMetals. Vol. 17.

6. **Boutamine Z, 2019.** Degradation of emerging pollutants in water by sonochemistry and advanced oxidation processes. s.l. PhD thesis Université BADJI MOKHTAR ANNABA.

7. **Eshghi Malayeri B, 1995.** decontamination of soils containing heavy metals using plants and microorganisms, . s.l.. D. thesis in organismal biology.

8. **Balet J.M, 2016.** Waste management. s.l. 5th Edition.

9. **Boubekri S, Affar F, 2014.** Localisation des décharges et dépotoirs sauvages, leur identification et leurs impacts sur l'environnement et la santé publique dans la commune de Bejaia. Bejaia: Mémoire de Master Université Abderrahmane mira de Bejaia, p 9.

10. **Adriano D, 1986.** Trace Elements in the Terrestrial Environment. Springer-Verlag, . NewYork :P 533. ISBN 978-1-4757-1909-3. DOI 10.1007/978-1-4757-1907-9.

11. **Baise D, 2000.** Teneurs Totales En " Métaux Lourds " Dans Les Sols Français Résultats Généraux Du Programme ASPITET. s.l. Journal de Courrier de l'environnement de l'INRA.

12. **Lisk D.J, 1988.** Environmental implications of incineration of municipal solid waste and ash disposal. s.l. Sci. Total. Environ.Vol. 74.

13. **Fethia A.M, and Mohamed S, 2000.** Mobility of heavy metals during the composting process. Tunisia : In: Water, waste, and environmental research.. p65.

14. **ADEME, 1990.** Knowledge and control of health aspects of sludge spreading.

15. **Mortvedt J. J. D, and Beaton J, 1995.** Heavy metal and radionuclide contaminants in phosphate fertilizers. New York : In: Tiessen H, editor. Phosphorus in the global environment: transfer, cycles and management pp 93-106.

16. **Jugurta, Ikkour, 2018.** Biotechnologie végétale et dépollution des sols, la phytoremédiation,. s.l. Mémoire de fin de cycle en vue d'obtention d'un diplôme de Master en sciences Agronomiques.

17. **Zorrig W, 2011.** Recherche et caractérisation de déterminants contrôlant l'accumulation de cadmium chez la laitue " Lactuca sativa". s.l. Thèse de doctorat. Montpellier SupAgro.

18. **Leteinturier B, 2002.** Evaluation du potentiel phytocenotique des gôts cuprifère d'Afrique Centro-australe en vue de la phytorémediation des sites pollués par l'activité minière, PhD thesis, Faculté Universitaire des sciences agronomiques de Gembloux,

19. **Chon H, Ahn J, Jung M.C,1998.** Seasonal variation and chemical forms of heavy metals in soil and dusts from the satellite cities of Seoul, Korea. Seoul : Environmental Geochemistry and Health, 20 :77-86.

20. **Lombi E, Zhao F.J, Dunham S.J and McGrat S.P, 2001.** Cadmium accumulation in populations of tlaspi caerulesces and Thlaspi goesingense. s.l. New phytologist 145:11-20.

21. **Farinet J, Niang S, 2004.** Le recyclage des déchets et effluents dans l'agriculture urbaine. In Développement Durable de l'Agriculture Urbaine en Afrique Francophone: Enjeux, Concepts et Méthodes, Montpelier: CIRAD, 143-172.

22. **Pasquini, M.W, Alexander, M.J,2004.** Chemical properties of urban waste ash produced by open burning on the Jos Plateau: implications for agriculture. s.l. The Science of the total environment, 319: 225-240.

23. **Adjia R, Fezeu W.M, Tchatchueng J.B, Sorho S, Echevarria G, Ngassoum M.B, 2008.** Long term effect of municipal solid waste amendment on soil heavy metal content of sites used for periurban agriculture in Ngaoundere, Cameroon. Ngaoundere : Afr. J. Environ. Sci. Techn,Vol. 2(12).

24. **Napoléon A et. al. 2023.** Impacts of uncontrolled landfills on soil quality in the peri-urban neighborhoods of the city of Goma (East DR Congo): the case of the MUGUNGA neighborhood. s.l. Int. J. Biol. Chem. Sci, 17(4): 1738-1749.

25. **Veron C,1990.** Mechanisms of contamination of the food chain by cadmium, principally the air-soil-plant-animal-human trophic chain. Annales des Falsifications, de l'Expertise chimique et Toxicologique. 83 (888): 201-224.

26. **Munro D.B, and Small E, 1998.** Vegetables of Canada.,. Canada: NRC Research Press,

27. **Madrpm, 2005.** Alternative crops: Quinoa, amaranth, spelt. Technology transfer in agriculture.

28. **Maughan P.J, Smith S.M, Fairbanks D.J, and Jellen E. N, 2011.** Development, characterization and linkage mapping of single nucleotide polymorphisms in amaranth (Amarantus spp) grains. Plant Genome. Vol. 4(1).

29. **Grubben G. J. H,2004.** Plant Resources of Tropical Africa 2. Vegetables. Wageningen,.

30. **AFNOR, 1981.** (Association Française de Normalisation),. Détermination du pH, NF ISO 10390. ln AFNOR qualité des sols, Paris, pp. 339-348.

31. **Black, Walkley A et al.1934.** An examination of the Detjareff method for deterrnining soil organic matter and a proposed modification of chromatic acid titration methhod. s.l. : soil science, 37: 29-38.

32. **Kjeldahl J, 1883.** A new method for the determination of Nitrogen in organic matter. s.l. Zeitschrift für analytische chemie 22, pp. 366-382.

33. **Bray R.H, and Kurtz L.T, 1945.** Determination of total organic and available forms of phosphorus in soils. s.l. : soil science , 59:39-45.

34. **Dommergues Y,1960.** La notion de coefficient de minéralisation du carbone dans les sols. . s.l. : l4agronomie Tropicale, vol15(1), pp 55-60.

35. **Baize D, 1994.** Total heavy metal content in French soils. First results from the ASPITET program. French: Courrier de l'Environnement de l'INRA,Vol. 22.

36. **Muller A, Homey B, Soto H, 2001.** Involvement of chemokine receptors in breast cancer metastasis. 410: 50-56.

37. **Salomon B, 2019.** Risques associés à l'usage des déchets urbains e agriculture péri-urbaine en Afrique Subsaharienne: impact sur les communautés bactériennes du sol et émergence de la multirésistance bactérienne. s.l. Thèse de doctorat unique université Joseph KI ZERBO.

38. **Bourgeon G, 1991.** Les sols rouges de l'indice péninsulaire méridionale: pédogenèse ferralitique sur socle cristallin en milieu tropical;. Paris : thèse de doctorat.

39. **Nhari F, Sbaa M, Vasel J.L, Fekhaoui M, 2014.** Soil contamination by heavy metals in an uncontrolled landfill: the case of the Ahfir- Saidia landfill (eastern Morocco). Morocco: J. Mater. Environ. Sci, Vol. 5.

40. **Brown S.L, Chaney R.L, Angle J.S, and Baker A.J.M, 1994.** Phytoremediation of Thlaspi caerulescens and Bladder campion for zinc-and cadmium-contaminated soil. . s.l. J. Environ. Quai. 23; 1151-1157.

41. **Mbouaka M.E, 2000.** Etude de l'efficacité agronomique des composts d'ordures ménagères au Burkina Faso : cas de la ville de Ouagadougou. . Ouagadougou: Mémoire d'ingénieur, Université Polytechnique bobo dioulasso .

42. **Mirsa R. V, Roy R. N, Hiraoka H,2005.** Farm-level composting methods. . s.l. Land and Water Working Papers 2. Food and Agriculture Organization of the United Nations, Italy, Rome, p. 35.

43. **Genot V, Colinet G, Bock L, 2007.** Fertility of agricultural and forest soils in the Walloon region,. s.l. FUSAG report, p. 75.

44. **Evariste A et al. 2015.** Measuring C02 fluxes and carbon sequestration in West African terrestrial ecosystems. s.l. : Biotechnol. Agron. Sc. Environ, 68-82.

45. **Joël M, 2020.** Spatio-temporal characterization of the amount of CO2 released by public landfills in the city of Kisangani (DR Congo). Kisangani.

46. **Ago E.E et al. 2015.** Carbon dioxide fluxes from a degraded woodland in West Africa and their Reponses to main environmental factors. Carbon Balance Manage.

47. **Juwarkar A. A, Yadav S. K, Kumar G. P and Singh S.K, 2008.** Effect of biosludge and biofertilizer amendment on growth of Jatropha curcas in heavy metal contaminated soils. s.l. Environmental Monitoring and Assessment, 145 : 7 - 15

48. **Alcaniz, Ortiz O, 2006.** Bioaccumulation of heavy metals in Dactylis glomerata L. growing in a calcareous soil amended with sewage sludge.

49. **Lingle J, Davis R,1959.** The influence of soil temperature phosphorus fertilization on the growth and mineral absorption of tomato seeding. s.l. J. Am. Soc. Hortic. Sci. 73: 312-322.

50. **Soltner D, 1986.** Les bases de la production végétale. s.l. 14th edition: collection sciences et techniques agricoles.

51.Tessier M, 2005. Quelques notions de fertilisation:. s.l. Conseil pour le développement de l'agriculture du Québec.

52.Bodjona B. M, Kili A. K, 2012. Tchegueni S., Kennou B., Tchangbedji G. Evaluation de la quantité des métaux lourds dans la décharge d'Agoè (Lomé-Togo): cas du plomb, cadmium, cuivre, nickel et zinc. Lomé: International Journal of Biological and Chemical Science.

53. **Blaise D, 2000.** Teneurs totales en " métaux lourds " dans les sols francais. Courrier de l'Environnement de l'INRA, Vol. 39.

54. **Mench M, Baize D, 2007.** Contamination des sols et de nos aliments d'origine végétale par les éléments en traces. Courrier de l'Environnement de l'INRA, Vol. 52.

55. **Kao T, El Mejahed K, Bouzidi A, 2007.** Assessment of metal pollution in agricultural soils irrigated by wastewater from the city of Settat (Morocco).

56. **Tankari dan-badjo A, Guéro Y, Dan Lamso N, Tidjani A.D, Ambouta J.M.K, Echevarria G, 2013.** Evaluation de la contamination des sols par les éléments traces métalliques dans les zones urbaines et périurbaines de la ville de Niamey (Niger). Niamey.

57. **Bowen H.J.M, 1979.** Environmental Chemistry of the Elements. New York: Academic Press, pp. 49-62.

58. **Soclo H, 1999** .Recherché de compost type et toxicité résiduelle au Bénin,. Benin: TSM 9(94)

59. **Adefemi S.O, Awokunmi E.E, 2009.** The impact of municipal solid waste disposal in Ado-Ekiti metropolis, Ekiti-State, Nigeria: Afr. J. Environ. Sci. Tech, Vol. 3(8).

60. **Larcher W,2003.** Physiological plant ecology. s.l. 4th ed.Springer.

61. **Ondo J.A, Biyogo R.M, Mebale A.A, Eba F, 2012.** Pot experiment of the uptake of metals by Amaranthus cruentus grown in artificially doped soils by copper and zinc. s.l. Food Science and Quality Management,Vol. 9.

62. **Kabata-Pendias A. and Pendias H, 2001.** Trace elements in soils and plants s.l. Third Edition. CRC Press. p 413.

63. **FAO/WHO, 2015.** General Standard for Contaminants and Toxins in Food and Feed. s.l.: CODEX STAN. CODEX STAN. p.64.

APPENDICES

Appendix1 : Soil sampling and characterization at the Saaba and Toudouwéogo landfills

Appendix2 : Vegetative growth and measurement of agro-morphological plant parameters

Appendix3 : Weighing of above-ground and root biomass

Printed by Books on Demand GmbH, Norderstedt / Germany